1 MONT

FREE
READING
at
www.ForgottenBooks.com

By purchasing this book you are eligible for one month membership to ForgottenBooks.com, giving you unlimited access to our entire collection of over 1,000,000 titles via our web site and mobile apps.

To claim your free month visit:
www.forgottenbooks.com/free1289557

English
Français
Deutsche
Italiano
Español
Português

www.forgottenbooks.com

Mythology Photography **Fiction**
Fishing Christianity **Art** Cooking
Essays Buddhism Freemasonry
Medicine **Biology** Music **Ancient
Egypt** Evolution Carpentry Physics
Dance Geology **Mathematics** Fitness
Shakespeare **Folklore** Yoga Marketing
Confidence Immortality Biographies
Poetry **Psychology** Witchcraft
Electronics Chemistry History **Law**
Accounting **Philosophy** Anthropology
Alchemy Drama Quantum Mechanics
Atheism Sexual Health **Ancient History**
Entrepreneurship Languages Sport
Paleontology Needlework Islam
Metaphysics Investment Archaeology
Parenting Statistics Criminology
Motivational

ISBN 978-0-656-32458-3
PIBN 11289557

For support please visit www.forgottenbooks.com

The Institute has attempted to obtain the best original copy available for filming. Features of this copy which may be bibliographically unique, which may alter any of the images in the reproduction, or which may significantly change the usual method of filming, are checked below.

L'Institut a microfilmé le meilleur exe lui a été possible de se procurer. Les d exemplaire qui sont peut-être uniques bibliographique, qui peuvent modifier reproduite, ou qui peuvent exiger une dans la méthode normale de filmage so ci-dessous.

[✓] Coloured covers/
Couverture de couleur

[✓] Covers damaged/
Couverture endommagée

[ ] Covers restored and/or laminated/
Couverture restaurée et/ou pelliculée

[ ] Cover title missing/
Le titre de couverture manque

[ ] Coloured maps/
Cartes géographiques en couleur

[ ] Coloured ink (i.e. other than blue or black)/
Encre de couleur (i.e. autre que bleue ou noire)

[ ] Coloured plates and/or illustrations/
Planches et/ou illustrations en couleur

[✓] Bound with other material/
Relié avec d'autres documents

[✓] Tight binding may cause shadows or distortion along interior margin/
La reliure serrée peut causer de l'ombre ou de la distorsion le long de la marge intérieure

[ ] Blank leaves added during restoration may appear within the text. Whenever possible, these have been omitted from filming/
Il se peut que certaines pages blanches ajoutées lors d'une restauration apparaissent dans le texte, mais, lorsque cela était possible, ces pages n'ont pas été filmées.

[ ] Coloured pages/
Pages de couleur

[✓] Pages damaged/
Pages endommagées

[ ] Pages restored and/or laminated/
Pages restaurées et/ou pelliculées

[✓] Pages discoloured, stained or fox
Pages décolorées, tachetées ou pi

[ ] Pages detached/
Pages détachées

[✓] Showthrough/
Transparence

[ ] Quality of print varies/
Qualité inégale de l'impression

[ ] Continuous pagination/
Pagination continue

[ ] Includes index(es)/
Comprend un (des) index

Title on header taken from:/
Le titre de l'en-tête provient:

[ ] Title page of issue/
Page de titre de la livraison

[ ] Caption of issue/
Titre de départ de la livraison

[ ] Masthead/
Générique (périodiques) de la liv

[✓] Additional comments:/
Commentaires supplémentaires: Pagination is as follows: p. 21-26.

This item is filmed at the reduction ratio checked below/
Ce document est filmé au taux de réduction indiqué ci-dessous.

| 10X | 14X | 18X | 22X | 26X |

The copy filmed here has been reproduced thanks to the generosity of:

University of Toronto Archives

The images appearing here are the best quality possible considering the condition and legibility of the original copy and in keeping with the filming contract specifications.

Original copies in printed paper covers are filmed beginning with the front cover and ending on the last page with a printed or illustrated impression, or the back cover when appropriate. All other original copies are filmed beginning on the first page with a printed or illustrated impression, and ending on the last page with a printed or illustrated impression.

The last recorded frame on each microfiche shall contain the symbol ➡ (meaning "CONTINUED"), or the symbol ▼ (meaning "END"), whichever applies.

Maps, plates, charts, etc., may be filmed at different reduction ratios. Those too large to be entirely included in one exposure are filmed beginning in the upper left hand corner, left to right and top to bottom, as many frames as required. The following diagrams illustrate the method:

1	2	3

1	2

4	5

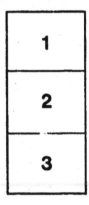

2   3

5   6

MICROCOPY RESOLUTION TEST CHART

(ANSI and ISO TEST CHART No. 2)

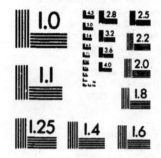

APPLIED IMAGE Inc
1653 East Main Street
Rochester, New York   14609   USA
(716) 482 - 0300 - Phone
(716) 288 - 5989 - Fax

UNIVERSITY OF TORONTO STUDIES

BIOLOGICAL SERIES

No. 7: AN EARLY ANADIDYMUS OF THE CHICK
BY R. RAMSAY WRIGHT

(REPRINTED FROM TRANSACTIONS OF THE ROYAL SOCIETY OF CANADA. 2ND SERIES. VOL. XI.)

THE UNIVERSITY LIBRARY: PUBLISHED BY
THE LIBRARIAN, 1907

University of Toronto Studies

COMMITTEE OF MANAGEMENT

Chairman: MAURICE HUTTON, M.A., LL.D.,
          Acting President of the University

PROFESSOR W. J. ALEXANDER, PH.D.

PROFESSOR W. H. ELLIS, M.A., M.B.

PROFESSOR A. KIRSCHMANN, PH.D.

PROFESSOR J. J. MACKENZIE, B.A.

PROFESSOR R. RAMSAY WRIGHT, M.A., B.Sc., LL.D.

PROFESSOR GEORGE M. WRONG, M.A.

General Editor: H. H. LANGTON, M.A.,
          Librarian of the University

III.—*An Early Anadidymus of the Chick.*

By PROFESSOR RAMSAY WRIGHT,

Biological Department, University of Toronto.

Read May 23rd, 1906.

The embryo which is described in the following pages was prepared and sectioned in June, 1905, for class purposes but its abnormality did not attract attention until it was brought into the laboratory. I am, therefore, unable to figure the surface view, and so far have not had leisure to model out its most interesting features.

The series contains 200 sections of 15 microns in thickness, corresponding to a length of 3 mm. in the hardened condition. The egg had been in the incubator for 24 hours, but, 10 somites having been observed, it was marked as practically equivalent in age to Duval's embryo of 29 hours (No. 1, Fig. 89 and Pl. XVI).

It was noted that the incubator was running at a temperature somewhat higher than the normal, which may account not only for its more rapid development but also for its abnormality, as may be inferred from Dareste (`   . 2, page 121).

Hertwig (No. 3:—   .l. I, p. 993) and others have remarked on the rarity of cases of Anadidymus in Sauropsida in comparison with the Ichthyopsida. This case is of particular interest, because, unlike Hoff-mann's (No. 4, page 40) there appears to be no indication of a double pr   tive streak, and, therefore, it is to be placed in the same category wit', .-areste's embryo (No. 2, Plate 16, Figs. 5 and 6), and possibly r .f Mitrophanow (whose paper I have not been able to consult)
    .estner (No. 5, page 88). The occurrence of such a case
        my opinion, invalidate the argument of Kaestner that
        are primitively double (No. 6, page 141), because it
            upon the degree, locality and method of the inter-
        two components, whether an organ shall appear double
        :   ure of section 131 (Fig. 13) would not be suspected
    iron an embryo otherwise than normal, while the inspection
s   126 (F   12) at once shows that each half of it in reality
        to a d .r   embryo. From this point, the interference
    has been   complete than cephalad, so that in the backward
_rowth of the r   tive streak region (cf. Hertwig, No. 2, pp. 895 and 896) the em_____   ears to be single.

Attention must be called to the contrast in the method of inter-ference in the head-region of my embryo and that in Kaestner's (No. 6, Taf. VII) where the ventral surfaces have interfered more than the dorsal, the result being a single heart and a double brain, instead of a double heart and a single brain (cf. my figure 9). The plane of interference becomes caudad more and more truly sagittal, so that the chordæ, at first widely divergent (Fig. 10), eventually fuse. (Fig. 13).

I now proceed to the description of the various systems of organs.

NERVOUS SYSTEM.

As a starting-point, I select section 12 (Fig. 5) through the region of the optic vesicles. It is easy to understand how the condition here pictured is arrived at we proceed from the normal state as s n in. Duval's Figs. 253 and 4. The two embryos have been inclined with their dorsal surfaces towards each other, and have interfered in such a way that the right and left lips of the neural groove of the one, have used with the right and left lips of that of the other. In this way, r om is left for the complete development of the " median " optic vesicles which, consequently, are very minute (ov'). The points of fusion are still noticeable and it is obvious that that of the left and right lips of the right and left components respectively (which now form the floor of the composite neural canal), is less complete, in such a way that some mesoderm cells have intruded into the neural canal at this point. The double character of the neural canal is brought strongly out by the two infundibula which diverge laterally towards the two blind foregut ends (ph.) beneath which the slightly thickened patches of ectoderm already indicate the hypophyses.

It is less easy to interpret the preceding sections (Figs.   to 4), but if two components such as are represented in Duval's Fig. 252 have interfered in such a way as materially to reduce in size the con-tiguous halves, then it becomes apparent that the convex floor of the composite neural canal in figure 4 is formed of the left and right brain-halves of the right and left components which have fused in the region of their dorsal neural sutures, while their ventral sutures are still widely separated. Still further forward (Fig. 3) these brain-halves are fused so that the most anterior end of the neural canal (Fig and 2) is formed of the lateral brain-halves only of the two components. It is noticeable that the separation of the brain from the ectoderm has apparently taken place sooner than is normal (No. 3, Vol. 2, page 252).

In the diencephalic region (Fig. 6) the brain is much compressed from side to side but it soon widens out into the mid-brain (Fig. 7).

In the trigeminal region of the hind-brain the neural canal is open for some thirteen sections, but before the auditory region is reached it is again closed as far as section 84, near which point (Fig. 11) there i again a failure to close for a few sections; thereafter, however, the canal is closed as far as sec 'on 126, Fig. 12, behind which point the groove is, at first narrowly, and then widely, open.

In section 160 (Fig. 16) the fusion of the ventral wall of the neural groove and the notochord begins and is continued in the following sections (Figs. 17-20), the complete fusion of the ectoderm, chorda, mesoderm and entoderm being attained at the 175th section (Fig. 20). Beyond this point we can hardly speak of a neural groove; the 181st section (Fig. 21), indeed, shows an unsymmetrical fissure which is not uncommon in the primitive groove of normal embryos, and by section 190 all traces of the primitive streak have disappeared and the germinal area presents a normal appearance (Fig. 23). The comparison of my Figures 15-22 with those of Hertwig (1 c., Figs. 536-545, page 891) shows that there is little difference except in the less amount of closure of the neural canal, and without an inspection of sections further forward, it would be impossible to detect any symptom of "duplicitas."

NOTOCHORD.

The conduct the two notochords has already been sufficiently referred to in the hinder region; it only remains to call attention to their gradual increase in size from their first appearance in section 9 (immediately behind figure 5) till their fusion in section 131, also to their gradual convergence to this point.

MESODERM.

As already remarked there are ten somites, and this is the case with the "median" series of fused somites which lie exactly in the same plane as the lateral ones. Of the "median" series, the seven posterior are better demarcated than those further forward, and are sometimes notched on their ventral surface. The rudiment of the Wolffian body may be seen in the region represented in Figs. 12 and 13.

VASCULAR SYSTEM.

A convenient starting-point for the description of the vascular system is the region depicted in Fig. 10 (section 67), where the vitelline veins are perfectly normal, and the only thing that arrests attention is the "median" descending aorta. Fig. 9 shows that the vitelline veins have not become fused into a single heart as in a normal embryo. Their endothelial tubes remain independent throughout, but the splanchnic mesoderm[1] does not at first dip in very far dorsad so as to furnish an independent wall for each heart. Further forward, however, it does so (Fig. 8), and eventually the two bulbs of the heart are widely separated and enclose between them a portion of the common cœlome (Fig. 7). But the two heart-tubes as seen in Fig. 9 do not contract gradually into the condition seen in Fig. 8; on the contrary, there is a marked constriction at the opening of each heart into its bulhus, beyond which a ventricular *cul-de-sac* extends cephalad for a few sections on each side.

The picture presented by Fig. 6 is best calculated to show the anterior duplicity of the vascular system, because when each bulbus approaches the stomatodœum it divides into two ventral aortæ. Of these the lateral aortæ alone form arches up the sides of the pharynx, for the median ones first anastomose below the pharynx, then subdivide into four small vessels which bend round its anterior surface, and finally open into the large vascular space represented in Fig. 5, situated between its anterior diverticula. Tracing this space backwards dorsad of the composite pharynx, we first find four vessels similar to those referred to above, which soon, however, fuse into the "median" dorsal aorta. This retains its size until we reach the segmented region of the embryo, in which it tends to be obliterated opposite the somites and to expand again intersomitically. The "lateral" dorsal aortæ conduct themselves as in a normal embryo, and the same may be said of the veins as far as they are developed.

ENTODERMIC TRACT.

Proceeding cephalad from Fig. 11 in which the median ridge formed of the median row of somites alone distinguishes this from the entoderm of a normal embryo we find nothing remarkable until about midway between Figs. 8 and 9, there the lateral pouches of the pharynx reach a little nearer the ectoderm in the region of the first gill-clefts, but a few sections further forward (Figs. 6 and 5) the two stamatodæa at once arrest attention, as do the two anterior diverticula corresponding to the pouches of Seosel of normal embryos.

I venture to enter a mild protest against Professor Kaestner's note (No. 6, p. 128) on the usage of the words somatopleure and spanchnopleure. Surely, if it is desirable to have mononyms for "somatic mesoblast," and "splanchnic mesoblast," it would be easy enough to form them instead of using terms which were invented and are constantly used to designate something else. If the language of anatomists knows only one meaning for πλευρα' that of zoologists is not so restricted. A Pleuronectid does not swim on its "pleura!"

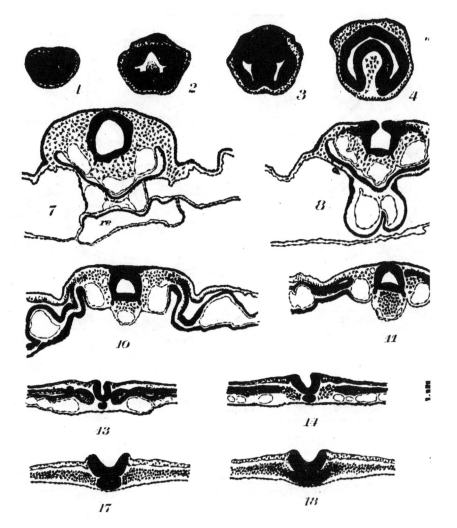

Trans. R. S. C., Sec. IV., 1906.

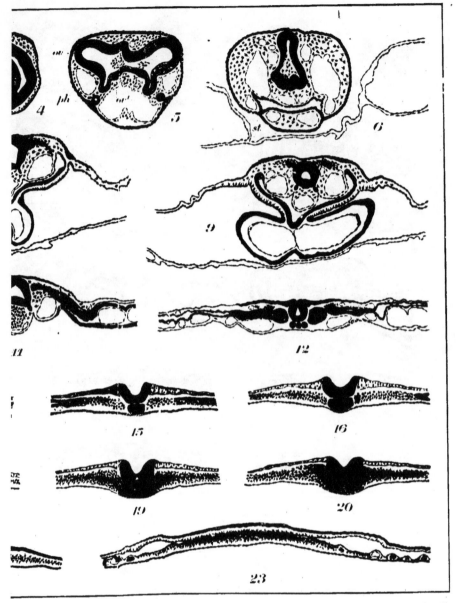

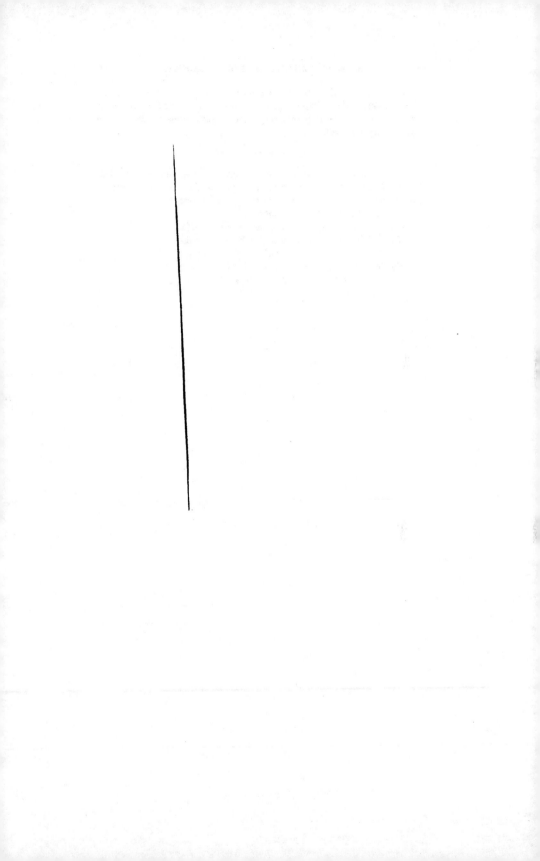

In conclusion, in spite of the apparent posterior simplicity of this embryo I am of the opinion that it can best be explained by assuming a double gastrulation at points very close to each other on the surface of the embryonic area.

LITERATURE CITED.

I have thought it unnecessary to cite all the papers consulted. Hertwig (No. 3) and Kaestner (No. 6) give a full list of papers to some of which, unfortunately, I have not had access.

No. 1.   Duval—Atlas d'Embryologie.
No. 2.   Dareste—-Production des Monstruosités.
No. 3.   Hertwig—Handbuch der Entwickelungslehre.
No. 4.   Hoffmann—Arch. mikr. Anat. XLI.
No. 5.   Kaestner, Arch. Anat. Phys., '98·
No. 6.   Kaestner, Arch. Anat. Phys., '02·

EXPLANATIONS OF THE FIGURES ON PLATE.

The sections were projected and carefully outlined on the drawing paper by means of the Zeiss Epidiascope and 20 mm. micro-planar, at such distances as to give an enlargement of 102 for figures 1 to 9, and 116 for figures 10 to 23.

Subsequently, the drawings, which were made by Mr. J. R. G. Murray, student in biology, University of Toronto, were reduced rather more than one-third, so that the magnification is respectively 63 and 72.

Figs. 1-4,— Nos. 4, 5, 6, and 8, of the series, through the fore-brain.

Fig. 5,— No. 12, through the anterior blind ends—ph.— of the pharynx. Ov. and ov' the right and left optic vesicles of the right component.

Fig. 6,— No. 19, through the stomatodæa of both components and the diencephalic region; round the composite pharynx are grouped eight arteries; two ventral, and two dorsal aortæ on each side.

Fig. 7.—No. 33, through the mesencephalon. Ventrad of the pharynx are the two aortic bulbs; dorsad, the median dorsal aortæ have united into a single vessel; re, ectodermic recess under the head.

Figs. 8, 9, and 10.-- Nos. 47, 55, and 67, respectively, through the fifth, seventh and eighth, and ninth nerves.

Fig. 11,— No. 80, through the second intersomite. The median dorsal aortæ have given place to a mass of mesoderm.

Fig. 12,— No. 126, behind the last somite. The chordæ are gaining in size, and the mesodermic mass diminishing. The rudiment of the Wolffian body is seen in this and in Fig. 13.

Fig. 13,— No. 131, the chordæ have fused.

Figs. 14 and 15,-- Nos. 150 and 154, the chorda and the wall of the neural groove gain in size.

Fig. 16,— No. 160, the beginning of the fusion between the floor of the neural groove and chorda.

Figs. 17, 18, 19 and 20,— Nos. 164, 168, 171 and 175, respectively, show the progressive fusion of the neural wall, chorda, mesoderm and entoderm.

Figs. 21 and 22,— Nos. 181 and 186, are through the hinder end of the primitive streak. The former shows traces of an oblique fissure.

Fig. 23,— No. 196, shows the nature of the mesoderm behind the primitive streak.

CPSIA information can be obtained
at www.ICGtesting.com
Printed in the USA
LVHW012239071118
596319LV00021B/311